NIST Calibration Services for Liquid Volume
NIST Special Publication 250-72

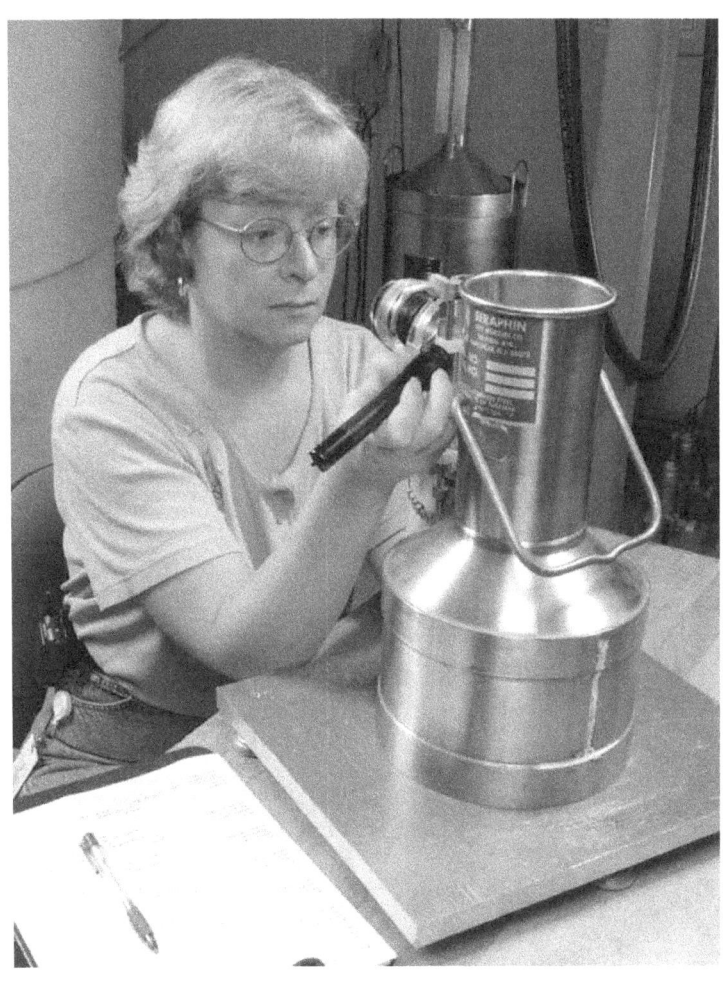

Vern E. Bean, Pedro I. Espina, John. D. Wright, John F. Houser, Sherry D. Sheckels, and Aaron N. Johnson
November 24, 2009

Fluid Metrology Group
Process Measurements Division
Chemical Science and Technology Laboratory
National Institute of Standards and Technology
Gaithersburg, Maryland, 20899

Table of Contents   Page

1. Description of Service. 1
2. Design Philosophy and Theory 1
3. Description of Facilities 4
    3.1 Gravimetric Facility 4
    3.2 Volume Transfer Facility 5
4. Calibration Procedures..5
    4.1   Gravimetric Methods...6
    4.1.1 Gravimetric Method: Direct Weighing of a Neck Scale Volume Provers..................6
    4.1.2 Gravimetric Method: Slicker Plate Volume Provers...8
    4.1.3 Gravimetric Transfer Method: Pre-weighed Water...9
    4.1.4 Gravimetric Method: Weigh Tank 10
    4.2   Volume Transfer Method 11
    4.2.1 Volume Transfer Method: Neck Scale Volume Prover to Neck Scale Volume
          Prover 11
    4.2.2 Volume Transfer Method: Slicker Plate Volume Prover to Neck Scale
          Volume Prover 14
    4.2.3 Volume Transfer Method: Slicker Plate Volume Prover, Delivered Volume 14
    4.3 Neck Scale Calibrations 15
5. Viscosity Corrections..16
6. Applying the Volume Calibration Results..17
7. Uncertainty..18
    7.1 Uncertainty of the Gravimetric Method 19
    7.2 Uncertainty of the Volume Transfer Method 27
8. References...31
Appendix A. Calibration of Weigh Scales..32
Appendix B. Derivation of Equations for Gravimetric Method for Neck
    Scale Provers...35
Appendix C. Derivation of Equation for the Volume Transfer Method...........................37
Appendix D. Sample Calibration Report..1

1. Description of Service.

The National Institute of Standards and Technology (NIST) provides calibration services for metal volume provers for volumes up to 2000 gal (7600 liters[†]). Volume provers with volumes up to 1900 L (500 gal) are normally calibrated gravimetrically; larger volumes are calibrated by volume transfer using provers that have been calibrated gravimetrically. The uncertainty in the volume calibration depends upon the method used as well as the design and volume of the prover. For slicker plate type volume provers, the uncertainty may be as low as 0.004 %. For neck scale type volume provers, the uncertainty is a function of the volume and neck diameter. Typical uncertainties for neck scale provers are 0.015 % + 0.003 gal or 0.015 % + 0.012 L [about 0.3 % for a 1 gal (3.8 L) volume prover and 0.02 % for a 100 gal (380 L) prover].

Customers should consult the web address http://ts.nist.gov/MeasurementServices/Calibrations/mechanical_index.cfm to find the most current information regarding our calibration services, calibration fees, technical contacts, turn around times, and instrument submittal procedures.

2. Design Philosophy and Theory

There are two common types of volume provers, distinguished by the way their full condition is determined; they are the slicker plate type and the neck scale type. For the slicker plate type, shown in Figure 1, the container is overfilled with water as permitted by the surface tension and then the excess water is sheared off with a flat transparent plate thus defining the volume as that enclosed by the container and the plate. Transparency of the plate enables inspection for voids (i.e., bubbles). For the neck scale type, shown in Figures 2 and 3, the volume is defined by reading a sight-glass-and-scale unit mounted on its neck. Either type of volume prover may be designed to be emptied by pouring out of the top (for the smaller sizes) or through a valve at the bottom. The top pouring style is shown in Figure 2 and the bottom drain type in Figure 3. More recommendations regarding the design of volume provers can be found in NIST Handbook 105-3 [Harris, 1997].

The horizontal cross section of any metal volume prover must be circular and its shape must permit complete emptying and draining. These volume provers must be sufficiently rigid to prevent appreciable distortion when full of water. The inside surface must be corrosion resistant and clean. All inside seams must be smooth to eliminate any traps for air (when being filled) or water (when being drained).

NIST maintains facilities for calibrating volume provers based on two different categories of methods: gravimetric methods and volume transfer methods. In the gravimetric method, the volume of distilled water required to fill a volume prover up to its full reference mark is determined by measuring the mass and the temperature of the water, calculating the density of the water from an appropriate empirical expression, and then calculating the volume from the definition of density. Distilled water is used because its density is well known as a function of temperature. The gravimetric volume calibration technique is a primary calibration method since it is directly traceable to NIST mass and temperature standards.

[†] The US gallon (231 in^3, 0.003785412 m^3, 3.785412 L) is the common unit of liquid volume used in commerce in the United States. Accordingly, to better communicate with the intended audience, in this publication US gallons are used in descriptions of volume provers for which the manufacturers have used US gallons, rather than the equivalent of gallons expressed in the less familiar m^3 of the International Systems of Units (SI). SI units are normally used in publications of the National Institute of Standards and Technology.

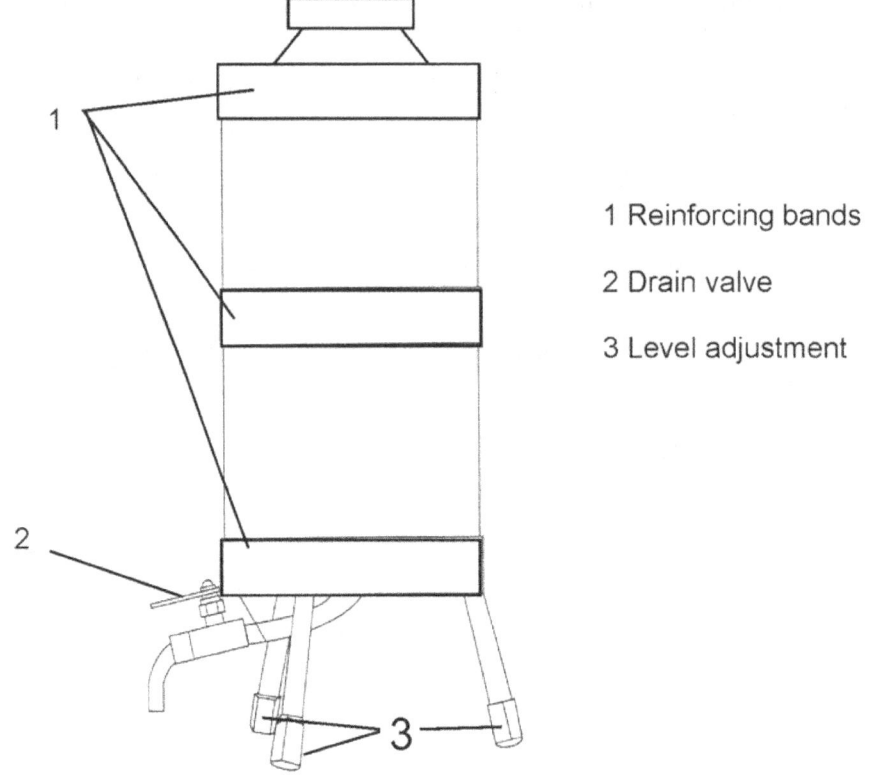

Figure 1. Schematic of slicker plate volume prover with a bottom drain. Top emptying styles are available in the smaller sizes.

In the volume transfer method, a volume prover is calibrated by filling it to its full reference mark with a known volume of tap water determined using working standard volume provers which were previously calibrated by the gravimetric method. Only the thermal expansion over a limited temperature range need be known for the water used for the calibration by transfer, rather than the density as in the gravimetric method. Measurements done by the NIST Fluid Metrology Group have shown that the difference in the thermal expansion of distilled water and tap water is negligible over the temperature range of interest and the use of tap water is far more convenient and less expensive than distilled water. The volume transfer method is used for calibrating volume provers that are too large to fit in the NIST gravimetric facility (usually in excess of 1900 L or 500 gal).

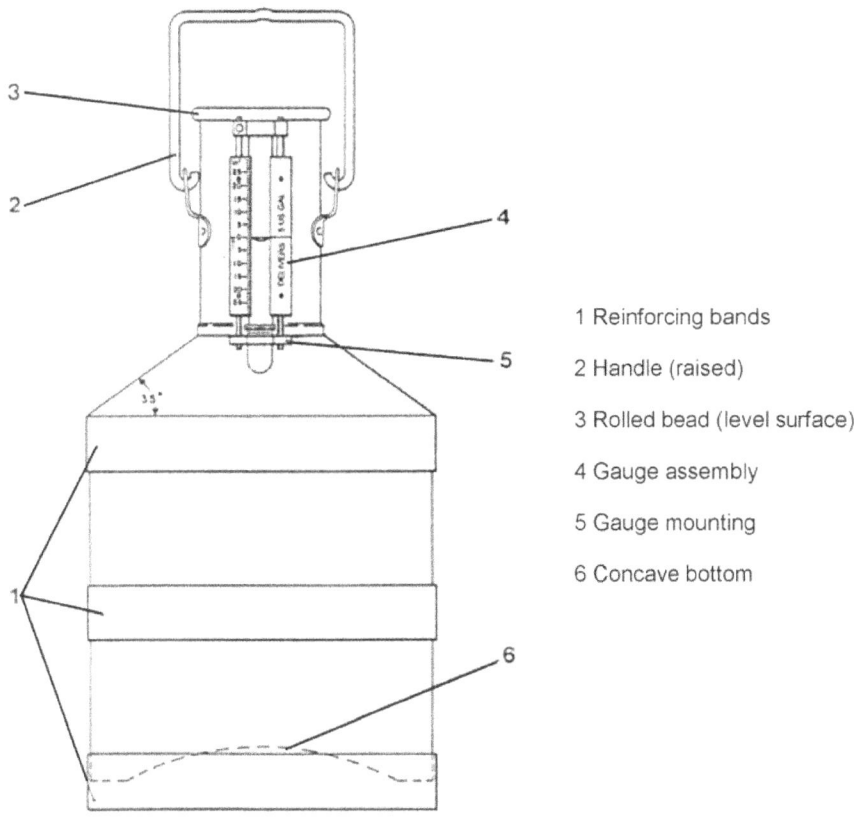

Figure 2. Neck scale volume prover, top emptying style.

1 Reinforcing bands
2 Handle (raised)
3 Rolled bead (level surface)
4 Gauge assembly
5 Gauge mounting
6 Concave bottom

Measurements of contained volume and delivered volume are both vital to commerce. <u>Contained volume</u> is the volume of liquid required to fill a dry volume prover to the full reference mark. <u>Delivered volume</u> is the volume of liquid that drains from a volume prover, originally filled to the full reference mark, under specified conditions. For a volume prover with a bottom drain, it is normally the volume that drains from the volume prover during the time from the opening of the valve to the closing of the valve 30 seconds after the cessation of the main flow. For a top pouring volume prover, it is normally the volume emptied from the volume prover from the time of first issue until 10 seconds after the cessation of the main flow with the volume prover held at an angle of 70° with respect to horizontal. Delivered volume is also the volume of water required to fill a volume prover to its full reference mark that has been pre-wetted by being filled to the full reference mark and then drained under the specified conditions. The contained volume is always larger by a few hundredths of a percent than the delivered volume for a given volume prover due to water that clings to the wall of the volume prover after draining.

To assess repeatability of the volume prover, and to check the quality of the calibration results, the methods described herein are normally repeated 5 times and the results analyzed statistically.

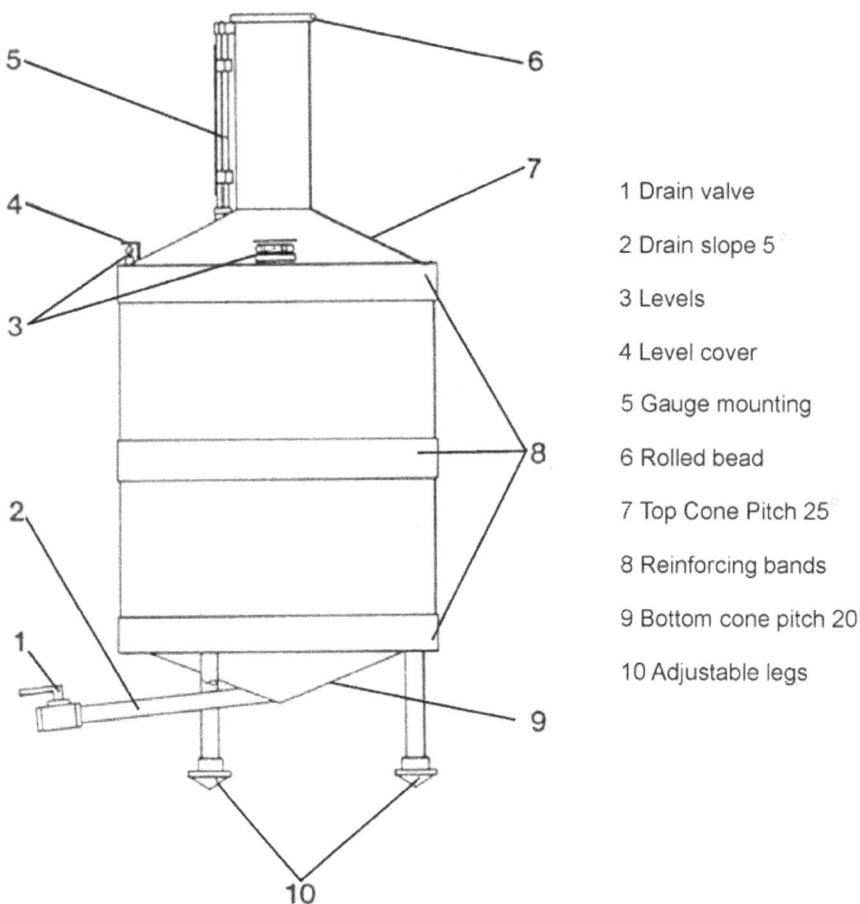

Figure 3. Neck scale volume prover, bottom drain style.

3. Description of NIST Facilities

3.1 Gravimetric Facility

The mass of distilled water required to fill a volume prover is determined using weigh scales. Since downward air currents > 0.15 m/s on the platform of a weigh scale can perturb the mass measurements, careful control of the direction and of the strength of air currents is important. The control of air currents, as well as the ambient temperature, is achieved by locating the gravimetric facility in an isolated laboratory module (a room-within-a-room) with its own ventilation and temperature control systems.

Two weigh scales, of 60 kg and 600 kg capacity, have been installed in the module. The 60 kg weigh scale is calibrated with weights from a 65 kg set which had been calibrated by the NIST Acoustics, Mass, and Vibrations Group. The 600 kg weigh scale is calibrated with a set of twelve 45 kg weights, which are calibrated on the 60 kg weigh scale. Both weigh scales are used in the up-loading cycle only to avoid the consequences of hysteresis. Further details of the calibration of the weigh scales are given in Appendix A.

The direct-weighing gravimetric method (see § 4.1.1 and § 4.1.2) is used for volume provers smaller than 380 L (100 gal). Typically, the 60 kg weigh scale is used for volume provers of 38 L (10 gal) or smaller, while the 600 kg weigh scale is used for volume provers from 38 L to 380 L (10 gal to 100 gal).

The floor area of the laboratory module is 4.9 m by 6.1 m, and its ceiling is 4.9 m high. In one corner is a mezzanine, 1.8 m by 2.4 m in area, located 2.1 m above the floor, which is designed to support the 600 kg weigh scale with a 456 L container of water such that weighed amounts of distilled water can be drained into volume provers that exceed the capacity of the weigh scale (see gravimetric transfer method, § 4.1.3). This permits gravimetric calibration of volume provers up to 1900 L, the maximum size that can be accomodated by the laboratory module.

The laboratory module is equipped with a 0.2 m diameter floor drain and a crane with a lifting capacity of 1000 kg. The module contains a 9500 L storage tank for distilled water with a plumbing system to distribute the water to 3 convenient sites within the module. Distilled water is supplied by a still located in the adjacent room.

Water and air temperatures are measured with thermistors traceable to the NIST Thermometry Group. Experiments have shown that the mixing during filling of the volume prover with water in a normal calibration cycle results in temperature gradients that are 40 mK or less and this is considered as a component in the uncertainty analysis.

Room temperature, barometric pressure, and room relative humidity measurements are made in order to make air buoyancy corrections to all mass measurements.

3.2 Volume Transfer Facility

The Fluid Metrology Group maintains a volume transfer facility that is used to calibrate volumes between 1900 L (500 gal) and 7600 L (2000 gal) by the volume transfer method (see § 4.2). This facility has a lift with a platform of sufficient capacity to support a filled 380 L (100 gal) volume prover at the appropriate height so as to drain into a volume prover under test located on the floor. Water temperature is measured with thermistors traceable to the NIST Thermometry Group.

4. Calibration Procedures

There are four calibration situations to which the gravimetric facility is applicable and three calibrations situations to which the transfer facility is applicable. Detailed procedures for all seven situations are presented for the sake of completeness, however not all of the methods are routinely applied in our calibration service.

All of the calibration methods share the following initial steps:

1. Inspect the volume prover for damage and/or rust. Ensure that the interior of walls of the prover (including the neck and sight glass if applicable) are clean and dry.

2. Record the manufacturer's label information: maker, model number, serial number, nominal volume, volumetric coefficient of thermal expansion, maximum and minimum neck scale values, and volume represented by each scale division (if applicable). Also, make a note describing the condition of the drain valve (e.g., new, old, rusted, etc).

Whenever measurements of the neck scale are made, appropriate scale reading aids are used to avoid errors due to parallax.

A bottom drain style volume prover (i.e., with a valve) is drained by opening the valve and leaving it open for 30 seconds after the cessation of the main flow. A top pouring style volume prover is drained by pouring, with the volume prover held at an angle of approximately 70° with respect to horizontal, including 10 seconds after the cessation of the main flow.

Generally, five measurements of the prover volume are made in order to assess the repeatability of the calibration procedures. The 5 repeated volume measurements also serve as a quality control measure: outliers can be detected (using Chauvenet's criterion [Coleman and Steele, 1999]) and replaced by new data sets.

With the exception of the weigh tank method, all of the methods can be used to determine contained or delivered volume. However, to measure the contained volume, the prover must initially be completely dry. So using the direct weighing method as an example, a single mass of the prover measured when it is dry is used with the 5 subsequent full mass measurements. This is done because it is impractical to dry the prover between repeated volume measurements, but it should be noted that the contained volume is not based on 5 separate dry mass measurements. In the case of volumetric transfer methods, the first filling of the dry prover delivers the contained volume. Five subsequent fillings give 5 measurements of the delivered volume.

In the case of provers with neck scales, the 5 repeated measurements are made close to the zero mark on the neck scale. During the last of the five repeats, the neck is filled to 5 different levels (one of which is nominally zero) in order to perform a neck scale calibration.

4.1 Gravimetric Methods

The four volume calibration situations to which the gravimetric volume facility is applicable are:

1. direct weighing of neck scale type volume provers to determine contained and/or delivered volume,
2. direct weighing of slicker plate type volume provers to determine contained and/or delivered volume,
3. pre-weighing the water before filling neck scale type volume provers to determine contained and/or delivered volume, used for the calibration of volume provers that exceed the limits of the 600 kg weigh scale, and
4. emptying a volume prover into a suitable container, which is weighed to determine the delivered volume of a volume prover, known as the weigh tank method.

All of the gravimetric methods require calibration of the weigh scale before its use via the procedure given in Appendix A.

Each of the four gravimetric methods has a separate procedure given below.

4.1.1 Gravimetric Method: Direct Weighing of a Neck Scale Volume Provers

The calibration of a neck scale type volume prover by the direct weighing consists of the following steps [Houser, 1973]:

1. Level the volume prover on the platform of the weigh scale.
2. Weigh the empty, clean, dry volume prover and record the weigh scale reading.
3. Fill the volume prover with distilled water up to an appropriate mark on the neck scale.
4. Inspect the volume prover for leaks.

5. Measure the temperature of the water in the volume prover, record the readings, and then remove the temperature sensors.
6. Measure and record the room air temperature, atmospheric pressure, and relative humidity.
7. Read and record the deviation of the location of the meniscus from zero on the neck scale, in scale divisions.
8. Weigh the filled volume prover and record the weigh scale reading.
9. Drain the volume prover.
10. Lift the volume prover off the weigh scale platform and restore it to nullify hysteresis effects.
11. Weigh the drained volume prover and record the weigh scale reading.
12. Calculate the air density, ρ_{air}, in kg/m^3, via [Jaeger and Davis, 1984]

$$\rho_{air} = \frac{3.4848 \times 10^{-3}}{(T + 273.15)} \left\{ P - 6.65287 \times 10^8 U \; e^{[-5315.56/(T + 273.15)]} \right\} \quad (1)$$

where:

P is the atmospheric pressure in Pa,
U is the relative humidity expressed in percent, and
T is the air average temperature in °C, all measured in step 6.

13. Calculate the indicated mass of the empty volume prover, m_e, via equation 3A of Appendix A.
14. Calculate the indicated mass of the filled volume prover, m_f, via equation 3A.
15. Calculate the indicated mass of the drained volume prover, m_d, via equation 3A.
16. Calculate the density of the distilled water, $\rho(T_w)$, in kg/m^3, via the equation of Patterson and Morris [1994]

$$\rho(T_w) = \rho_0 \left\{ 1 - \left[A(T_w - T_0) + B(T_w - T_0)^2 + C(T_w - T_0)^3 + D(T_w - T_0)^4 + E(T_w - T_0)^5 \right] \right\} \quad (2)$$

where:

ρ_0 = 999.97358 kg/m^3
A = 7.0134 x 10^{-8} (°C)$^{-1}$
B = 7.926504 x 10^{-6} (°C)$^{-2}$
C = -7.575677 x 10^{-8} (°C)$^{-3}$
D = 7.314894 x 10^{-10} (°C)$^{-4}$
E = -3.596458 x 10^{-12} (°C)$^{-5}$
T_0 = 3.9818 °C
T_w is the average temperature of the water in °C.

17. Calculate the contained volume of the volume prover in m^3 at T_w, adjusted to a neck scale reading of zero, via

$$V_{\text{contain}}(T_w) = \frac{m_f - m_e}{\rho(T_w) - \rho_{\text{air}}} - h \qquad (3)$$

where:

h is the deviation of the location of the meniscus from zero (in m³) with the appropriate algebraic sign applied. Values of h above zero are positive; those below zero are negative.

Equation (3) is derived in Appendix B.

18. Calculate the delivered volume of the volume prover in m³ at the temperature of the water during the test, T_w, adjusted to a neck scale reading of zero, via

$$V_{\text{deliver}}(T_w) = \frac{m_f - m_d}{\rho(T_w) - \rho_{\text{air}}} - h \qquad (4)$$

Equation (4) is derived in Appendix B.

19. Calculate the contained and delivered volumes of the prover at the desired reference temperature requested by the customer, via

$$V(T_{\text{ref}}) = C\, V(T_w)\left[1 - \beta_t (T_w - T_{\text{ref}})\right] \qquad (5)$$

where:

V is either the contained or delivered volume,
C is the appropriate conversion factor from m³ to the volume unit requested by the customer,
β_t is the volumetric coefficient of thermal expansion for the volume prover obtained from the volume prover label or from a handbook based on the volume prover material of construction, and
T_{ref} is the reference temperature specified by the customer. It is normally 15.56 °C (60 °F).

The volumetric coefficient of thermal expansion for the prover is 3 × the linear coefficient of thermal expansion. For the most common material of construction, stainless steel, we use a volumetric coefficient of thermal expansion of 48 parts in 10^6 / °C. Hence for a 10 °C increase in temperature, the contained and delivered volumes for a stainless steel prover will increase by 10 × 0.000048 = 0.048 %.

4.1.2 Gravimetric Method: Slicker Plate Volume Provers

The calibration of a slicker plate volume prover by the gravimetric method is done by modifying the procedure for the neck scale type volume prover:

Fill the slicker plate volume prover with distilled water within a few millimeters of the top of the neck. Measure the temperature of the water in the volume prover and record. Finish filling the volume prover such that the water extends above the top of the neck of the volume prover due to surface tension. Shear off the excess by sliding a transparent plate across the top of the neck. Absorb the excess water with towels and discard. Transparency of the slicker plate enables visual inspection for voids under the plate; do not allow bubbles. For a slicker plate volume prover, there is no neck scale and so $h \equiv 0$.

Omit steps related to measuring the height of the water in the neck and set h equal to 0 in equations 3 and 4.

4.1.3 Gravimetric Transfer Method: Pre-weighed Water

The pre-weighed water method is used to calibrate neck scale volume provers that have such large dimensions that they cannot be placed on the platform of the 600 kg weigh scale or have such capacity that the mass of the volume prover filled with water exceeds the capacity of the 600 kg weigh scale. The method is useful for volume provers with capacities up to 1900 L (500 gal), which is the maximum size that will fit in the volume laboratory module.

It is possible to obtain contained or delivered volumes via this method depending upon whether the volume prover is dry or pre-wetted at the outset of the calibration.

The procedures are:

1. Install the 600 kg weigh scale on the mezzanine and calibrate it with reference masses after moving (see Appendix A). Place the 120 gal (456 L) tank on the weigh sale in approximately the center of the scale pan.

2. Level the volume prover under test at an appropriate location on the floor so that it can be filled with successive volume transfers from the tank on the mezzanine.

3. Provide appropriate plumbing from the drain valve of the 120 gal tank on the weigh scale to the neck of the volume prover under test.

4. If the volume prover under test is to be calibrated for contained volume, the inside must be dry at the outset of the calibration. If it is to be calibrated for delivered volume, the inside must be pre-wetted at the outset which is accomplished by filling the volume prover with water up to the full reference mark, allowing the volume prover to drain through the valve, and closing the valve 30 seconds after the cessation of the main flow.

5. Fill the 120 gal tank with distilled water.

6. Record the weigh scale reading for the filled 120 gal tank.

7. Measure and record the room air temperature, atmospheric pressure, and relative humidity.

8. Drain the 120 gal tank into the volume prover under test without the loss of any water. It is advisable to place a finger over the opening of the sight glass tube of the volume prover under test to prevent water from jetting up the tube. This procedure is followed without interruption and completed within 3 hours so that loss by evaporation can be considered negligible.

9. Lift the 120 gal tank off the weigh scale platform and restore it to nullify hysteresis effects.

10. Record the weigh scale reading for the empty 120 gal tank.

11. Repeat steps 5 through 10 until the volume prover under test is filled up to the proximity of the zero mark on the neck scale.

12. Inspect the volume prover under test for leaks.

13. Read and record the deviation of the location of the meniscus from zero on the neck scale in volume units.

14. Measure the temperature of the water in the volume prover under test and record.

15. Calculate the air density, ρ_{air}, in kg/m^3, via equation 1.

16. Calculate the indicated masses of the filled 120 gal tank, m_{1i}, where i is the index tracking the tank refills, via equation 3A of Appendix A.

17. Calculate the indicated masses of the drained 120 gal tank, m_{2i}, where i is the index tracking the tank refills, via equation 3A.

18. Calculate the density of the distilled water, $\rho(T_w)$, in kg/m^3, via equation 2.

19. Calculate the volume in m^3, adjusted to a neck scale reading of zero, from

$$V_{cal}(T_w) = \sum_i \left[\frac{(m_{1i} - m_{2i})}{\rho(T_w) - \rho_{air}} \right] - h \quad (6)$$

where:

$V_{cal}(T_w)$ is either the contained volume or the delivered volume depending upon whether the volume prover was dry or pre-wetted at the outset of the calibration, in m^3,

m_{1i} is the indicated mass of the 120 gal tank and the water before draining into the volume prover under test, in kg, and

m_{2i} is the indicated mass of the 120 gal tank and the water after draining into the volume prover under test, in kg.

20. Calculate the contained and delivered volume at the desired reference temperature requested by the customer via equation 5.

4.1.4 Gravimetric Method: Weigh Tank

The weigh tank technique is used to determine the delivered volume of a prover by filling it with distilled water up to its full reference mark and emptying it into a suitable container that is weighed with and without the water. The weigh tank is usually a light-weight plastic or glass container. The technique is used when the mass of the volume prover filled with water exceeds the capacity of the weigh scale but the mass of the plastic tank and the water is within the capacity of the weigh scale. It is also used for "cubic foot bottles". The procedure is:

1. Level the volume prover under test at an appropriate elevation above the floor.

2. Provide appropriate plumbing from the drain valve of the volume prover under test to the weigh tank.

3. Fill the volume prover under test up to the full reference mark with distilled water.

4. Inspect the volume prover under test for leaks.

5. Weigh and record the weigh scale reading for the empty weigh tank.

6. Measure and record the room air temperature, atmospheric pressure, and relative humidity.
7. If possible, measure the temperature of the water in the volume prover under test and record.
11. Drain the volume prover under test into the weigh tank, following normal draining procedures.
12. Weigh and record the weigh scale reading for the weigh tank with the water.
13. If it was not possible to measure the temperature of the water in the volume under test, measure the temperature of the water in the weigh tank and record.
14. Calculate the air density, ρ_{air}, in kg/m³, via equation 1.
15. Calculate the indicated mass of the empty weigh tank, m_e, via equation 3A of Appendix A.
16. Calculate the indicated mass of the weigh tank and the water, m_f, via equation 3A.
17. Calculate the density of the distilled water, $\rho(T_w)$, in kg/m³, via equation 2.
18. Calculate the delivered volume at T_{ref} in m³ from

$$V_{deliver}(T_{ref}) = \left[\frac{m_f - m_e}{\rho(T_w) - \rho_{air}} \right] \left[1 - \beta_t (T_w - T_{ref}) \right] \tag{7}$$

where:

m_f is the mass of the weigh tank and water, in kg, and
m_e is the mass of the empty weigh tank, in kg.

4.2 Volume Transfer Method

Volumes larger than 500 gal and no larger than 2000 gal are calibrated by one of three transfer methods:

1. calibrate a neck scale volume prover for contained or delivered volume using a neck scale volume prover as the working standard,
2. calibrate a neck scale volume prover for contained or delivered volume using a slicker plate volume prover as the working standard, and
3. calibrate a slicker plate volume prover for delivered volume using a neck scale volume prover as the working standard.

It is not possible to calibrate a slicker plate volume prover for contained volume using the transfer method because the process of shearing off the water at the neck to define the full condition discards an unknown volume of the calibrated volume of water delivered by the working standard.

4.2.1 Volume Transfer Method: Neck Scale Volume Prover to Neck Scale Volume Prover

The calibration of a neck scale volume prover, using a neck scale volume prover for the working standard, consists of the following steps, assuming, for simplicity, emptying only once from one working standard is required [Houser, 1973]:

1. Place and level an appropriate working standard volume prover, the volume of which has been previously determined by gravimetric method, on the elevated platform set at the appropriate

height.

2. Level the volume prover under test at an appropriate location with respect to the elevated platform.
3. Provide appropriate plumbing between the drain of the working standard and the neck of the volume prover under test.
4. If the volume prover under test is to be calibrated for contained volume, the inside must be dry at the outset of the calibration. If it is to be calibrated for delivered volume, the inside must be pre-wet at the outset.
5. Fill the working standard with tap water up to a convenient mark on the neck scale.
6. Measure the temperature of the water in the working standard, record the readings, and remove the temperature sensors.
7. Read and record the deviation from zero of the location of the meniscus on the neck scale for the working standard in volume units.
8. Drain the working standard, without the loss of any water, into the volume prover under test, following normal drain procedures. It is advisable to place a finger over the opening of the sight glass tube of the volume prover under test to prevent water from jetting up the tube.
9. Read and record the deviation from zero of the location of the meniscus on the neck scale for the volume prover under test in volume units.
11. Measure the temperature of the water in the volume prover under test and record.
12. Calculate the density of the water in the working standard, $\rho(T_{w,s})$ via equation 2.
13. Calculate the density of the water in the volume prover under test, $\rho(T_{w,t})$ via equation 2.
14. Calculate the volume of the prover under test, $V_t(T_{w,t})$, adjusted to a neck sale reading of zero via,

$$V_t(T_{w,t}) = \frac{\rho(T_{w,s})\,(V_s + h_s)[1 + \beta_s(T_{w,s} - T_{ref,s})]}{\rho(T_{w,t})} - h_t \qquad (8)$$

where:

V_s is the delivered volume of the working standard at the reference temperature,

h_s is the deviation of the location of the meniscus from zero on the neck scale of the working standard, in volume units, with the appropriate algebraic sign applied,

β_s is the volumetric thermal expansion coefficient for the working standard,

$T_{w,s}$ is the temperature of the water in the working standard,

$T_{ref,s}$ is the reference temperature for the working standard,

h_t is the deviation of the location of the meniscus from zero on the neck scale of the volume prover under test, in volume units, with the appropriate algebraic sign applied,

$T_{w,t}$ is the temperature of the water in the volume prover under test.

Equation 8 will yield contained volume at $T_{w,t}$ if the inside walls of the prover under test were dry at the outset of the calibration, or delivered volume at $T_{w,t}$ if the inside walls of the prover were pre-wetted at the outset of the calibration. Equation 8 is derived in Appendix C.

Equation 8 shows why tap water can be used rather than distilled water in the volume-by-transfer calibration process. Equation 8 contains the ratio $\rho(T_{w,s})/\rho(T_{w,t})$. Following the form of equation 2, this ratio can be expressed as,

$$\frac{\rho(T_{w,s})}{\rho(T_{w,t})} = \frac{\rho(T_0)f(T_{w,s}-T_0)}{\rho(T_0)f(T_{w,t}-T_0)} = \frac{f(T_{w,s}-T_0)}{f(T_{w,t}-T_0)} \tag{9}$$

where:

$\rho(T_0)$ is the density at a specified temperature and is a constant,

$f(T_{w,s}-T_0)$ and $f(T_{w,t}-T_0)$ are the thermal expansion terms for the tap water at $T_{w,s}$ and $T_{w,t}$ respectively.

Measurements by the NIST Fluid Metrology Group have demonstrated that the thermal expansion coefficients A through E and T_0 of equation 2 also apply to tap water over the temperature range of interest. Hence, equation 2 can be used to calculate water density in equation 8 when tap water is used.

In some cases, calibration by transfer may require several emptyings of more than one working standard. Equation 8 requires the following modifications for this situation

$$V_t(T_{w,t}) = \frac{\sum_i [\rho(T_{w,s,i})(V_{s,i}+h_{s,i})[1+\beta_{s,i}(T_{w,s,i}-T_{ref,s})]]}{\rho(T_{w,t})} - h_t \tag{10}$$

where $\rho(T_{w,s,i})$, $h_{s,i}$, and $T_{w,s,i}$ may change when the same working standard is used for multiple emptyings, and, in addition, $V_{s,i}$ and $\beta_{s,i}$, will vary if more than one working standard is used. The index i is used to track the number of emptyings.

15. Calculate $V_t(T_{ref})$:

$$V_t(T_{ref}) = \frac{\rho(T_{w,s})(V_s+h_s)[1+\beta_s(T_{w,s}-T_{ref,s})]}{\rho(T_{w,t})[1+\beta_t(T_{w,t}-T_{ref})]} - h_t \tag{8'}$$

$$V_t(T_{ref}) = \frac{\sum_i [\rho(T_{w,s,i})(V_{s,i}+h_{s,i})[1+\beta_{s,i}(T_{w,s,i}-T_{ref,s})]]}{\rho(T_{w,t})[1+\beta_t(T_{w,t}-T_{ref})]} - h_t. \tag{10'}$$

4.2.2 Volume Transfer Method: Slicker Plate Volume Prover to Neck Scale Volume Prover

When a neck scale volume prover is calibrated using a slicker plate working standard, the procedure of Section 4.2.1. is followed with some modifications.

Fill the slicker plate volume prover with distilled water within a few millimeters of the top of the neck.

Measure the temperature of the water in the volume prover and record. Finish filling the volume prover such that the water extends above the top of the neck of the volume prover due to surface tension. Shear off the excess by sliding a transparent plate across the top of the neck. Absorb the excess water with towels and discard. Transparency of the slicker plate enables visual inspection for voids under the plate; voids must not be allowed to occur.

Omit steps involving reading the level of the water in the neck scale of the working standard and set $h_s \equiv 0$ in equations 8 and 10.

4.2.3 Volume Transfer Method: Slicker Plate Volume Prover, Delivered Volume

It is possible to calibrate the delivered volume of a slicker plate volume prover by emptying it into a neck scale volume prover, which has been calibrated for contained volume, provided one emptying of the slicker plate volume prover will fill the neck scale volume prover up into the neck region such that the neck scale can be read. The procedure is:

1. Place and level the volume prover under test on the elevated platform.

2. Provide appropriate plumbing between the drain of the volume prover under test and the neck of the working standard volume prover.

3. Level the working standard volume prover at an appropriate location with respect to the elevated platform.

4. Fill the volume prover under test with tap water within a few millimeters of the top.

5. Measure the temperature of the water in the volume prover under test and record.

6. Finish filling the volume prover under test to the limit permitted by the surface tension and shear off the excess water with a transparent plate.

7. Drain the volume prover under test, without the loss of any water, into the working standard, following normal draining procedures. It is advisable to place a finger over the opening of the sight glass tube of the volume prover under test to prevent water from jetting up the tube.

9. Read and record the deviation from zero of the location of the meniscus on the neck scale for the working standard, in volume units.

10. Measure the temperature of the water in the working standard and record.

11. Calculate the density of the water in the working standard, $\rho_s(T_{w,s})$, via equation 2.

12. Calculate the density of the water in the volume prover under test, $\rho_t(T_{w,t})$, via equation 2.

13. Calculate the delivered volume of the volume prover under test, $V_{deliver}(T_{w,t})$, via

$$V_{\text{deliver}}(T_{w,t}) = \frac{\rho(T_{w,s})(V_s + h_s)[1 + \beta_s(T_{w,s} - T_{\text{ref},s})]}{\rho(T_{w,t})} \quad (11)$$

4.3 Neck Scale Calibrations

Neck scales can be calibrated by various methods, including:

1. *Volumetric Method:* add known volumes of water to the prover under test and observe the change in neck scale reading,
2. *Displacement Method:* insert objects of known volumes (such as metal spheres of known diameter) through the neck into the prover and observe the change in neck scale reading,
3. *Gravimetric Method:* add water to the prover to attain the desired neck scale readings and observe the changes in mass of the filled prover with a weigh scale.

A more detailed description of the gravimetric method follows.

1. Calibrate the appropriate weigh scale, following the procedures given in Appendix A.
2. Determine which neck scale increments will be calibrated. For example, for a neck scale that ranges from -100 in^3 to 100 in^3, one might choose to calibrate the -80 in^3, -40 in^3, 0 in^3, 40 in^3, and 80 in^3 levels.
3. Fill the volume prover with water up to the lowest neck scale reading to be calibrated.
4. Read and record the deviation of the location of the meniscus from zero on the neck scale.
5. Record the mass indicated by the weigh scale.
6. Add distilled water to the volume prover under test such that the water level rises to the next neck scale set point, read and record the meniscus level on the neck scale, record the mass, and repeat this process until all the set points have been measured.
7. Measure the temperature of the water and record the value.
8. Measure the room air temperature, barometric pressure, and the relative humidity and record the values.
9. Drain the prover. Measure and record the drained weight.
10. Use equations 4 and 5, with h set equal to zero, to calculate the volume of water in the prover from the mass measurements. The correct neck scale reading can be calculated via

$$V_{\text{neck}} = V_{\text{deliver}}(T_{\text{ref}}) - V_{\text{nominal}} \quad (12)$$

where:

V_{nominal} is the nominal volume of the prover at a neck scale reading of zero. The calculated neck volume V_{neck} can be converted to the desired number of volume units (including the number of neck scale divisions).

Note: if the dry weight of the prover is used instead of the drained weight (equations 3 and 5), the neck scale calibrations will apply to contained volume rather than delivered volume.

5. Viscosity Corrections

Volume calibration is done at ambient laboratory temperatures, normally in the range of 21 °C to 24 °C. Usually, the customer requests that the volume be referenced to some other temperature (often it is 15.56 °C or 60 °F). The calculation of contained volume at the reference temperature is done using the thermal expansion coefficient of the material from which the volume prover is made, as expressed in equation 5. Gravimetric calibrations that we have conducted at room temperature and 15.5 °C have demonstrated this procedure to be correct.

Delivered volume is also a function of the viscosity of the water used. At higher temperatures, the viscosity will be lower and less water will cling to the interior walls of the prover. Thus, more water will drain out during the 30 s drain time, making the delivered volume at room temperature slightly larger than at 15.56 °C.

Theory suggests the difference between contained and delivered volume can be expressed as: [Bird, Stewart, and Lightfoot, 1960]

$$V_{contain}(T_w) - V_{deliver}(T_w) = kG\left[\frac{\nu(T_w)}{t}\right]^{1/2} \tag{13}$$

where:
- k is a proportionality constant,
- G is a factor due to the geometry of the volume prover,
- t is the 30 second drain time between the cessation of the main flow and the closure of the drain valve,
- $\nu(T_w)$ is the kinematic viscosity of water.

Similarly, at T_{ref},

$$V_{contain}(T_{ref}) - V_{deliver}(T_{ref}) = kG\left[\frac{\nu(T_{ref})}{t}\right]^{1/2} \tag{14}$$

where:
- $\nu(T_{ref})$ is the kinematic viscosity of water at the reference temperature.

Solving equations 13 and 14 for $V_{deliver}(T_{ref})$:

$$V_{deliver}(T_{ref}) = V_{contain}(T_{ref}) - [V_{contain}(T_w) - V_{deliver}(T_w)]\left[\frac{\nu(T_{ref})}{\nu(T_w)}\right]^{1/2} = V_{contain}(T_{ref}) - V_\nu \tag{15}$$

where:
- V_ν is the viscosity correction with units of volume.

The kinematic viscosity of water (dynamic viscosity divided by density) in m²/s is given by:

$$v(T) = 1.76263 \times 10^{-6} - 5.4994 \times 10^{-8} T + 1.04326 \times 10^{-9} T^2 - 9.178 \times 10^{-12} T^3 \tag{16}$$

Equation 16 was obtained by fitting a polynomial to the ratio of dynamic viscosity to density for water, over the temperature range from 10 °C to 30 °C, with values taken from the literature [Kell, CRC].

The viscosity correction is proportional to the difference between the contained and delivered volumes. This difference $V_{contain}(T_{ref}) - V_{deliver}(T_{ref})$ depends on the size of the prover. Based on results of volume determinations from 80 provers made at an average temperature of 22 °C, $V_{contain}(T_{ref}) - V_{deliver}(T_{ref})$ is approximately 0.02 % for a 100 gal prover and the difference increases to about 0.05 % for a 5 gal prover. (This is due to the surface area per unit volume being greater for a small prover than for a large prover.)

The viscosity correction in equation 15, $[v(T_{ref})/v(T_w)]^{1/2}$, is approximately 1.013 for a 1 °C temperature change near room temperature. Therefore, using a 100 gal prover at 22 °C as an example, we can expect it to deliver (22 °C – 15.56 °C) × 0.013/°C × 0.0002 = 0.000017 or 17 parts in 10^6 more liquid due to a smaller viscosity than it would if it were used at 15.56 °C. For a 5 gal prover, the same temperature difference should lead to:

$$(22\ °C - 15.56\ °C) \times 0.013/°C \times 0.0005 = 0.000048$$

or 48 parts in 10^6 increase in delivered volume.

Our data reduction spreadsheets list the uncorrected, as well as, the viscosity corrected values of the delivered volume. Unless otherwise requested by the customer, the text of our calibration report gives only the delivered volume without the viscosity correction and the viscosity correction is treated as an uncertainty component of the volume calibration.

Note that the weigh tank method does not provide enough data to determine the contained volume and hence the viscosity correction cannot be applied to this method.

6. Applying the Volume Calibration Results

Most of the volume provers calibrated at NIST have neck scales. Generally, there are 5 repeated volume determinations made at a nominal neck scale reading of zero and a neck scale calibration process that leads to calibration results at 4 set points spread over the range of the neck scale. Samples of results of this typical calibration are given in Tables 1, 2, and 3 for three different sized provers. Table 4 shows results for a 5 gal slicker plate type prover.

For many customers, the volume calibration is performed to demonstrate that the volume ascribed to the prover is the same as that determined at NIST, within some tolerance. But there are many other possible approaches for applying the calibration results when the prover is subsequently used as a working standard to measure volume. For instance, the calibration allows one to compare the volume determined by NIST to the volume indicated by the prover and to calculate a volume correction,

$$V_{corr}(T_{ref}) = V_{delivered}(T_{ref}) - V_{indicated}(T_{ref}). \tag{17}$$

In this case, we are determining the correction for delivered volume at the reference temperature. For the most common calibration performed, there are 9 values for V_{corr}, 5 at approximately zero on neck scale and 4 distributed over the range of the neck scale. One can calculate a first order least squares best fit on the 9 corrections in order to obtain the volume correction as a function of the neck scale reading and examples of this approach are shown in Tables 1 through 3. The zeroth order coefficient (intercept) of the linear fit can be added to the nominal prover volume to obtain the delivered volume as calibrated by NIST at a neck scale reading of zero. The first order coefficient (slope) of the best fit line can be used to correct for errors in the neck scale caused by mismatched neck cross sectional area and the length of a scale division.

Using Table 1 as an example, for a zero neck scale reading, the intercept value of -1.67 in^3 (= -4.18 divisions × 0.4 in^3 / division) should be added to the nominal volume of 9702 in^3 (42 gal) to obtain the NIST measured volume of 9700.33 in^3. If the prover is used at a neck scale reading other than zero, the corrected volume of the prover would be calculated via

$$V_{prover} = V_{nominal} + [a_0 + a_1 N_D V_{div}] \tag{18}$$

where a_0 and a_1 are the best fit intercept and slope respectively, N_D is the number of scale divisions relative to the neck scale zero (below zero is a negative value), and V_{div} is the volume represented by a single scale division. Alternatively, for a given neck scale reading, the correction (in scale divisions) can be read from a graph as presented in Table 1 or a data table generated by the user.

If a customer chooses to apply viscosity corrections, he should start with the viscosity corrected delivered volume given in the data reduction spreadsheet. Then equation 15 must be rearranged algebraically to give $V_{deliver}(T_w)$ alone on the left hand side. The resulting equation can be used to calculate the delivered volume with viscosity effects taken into account at the temperature conditions that the customer users the prover. If the customer uses the prover with a liquid other than water, the appropriate values for density and kinematic viscosity should be used in any version of equation 15.

7. Uncertainty

In the following sections we will quantify the uncertainty of the volume calibration results for the methods most commonly used at NIST: 1) direct weighing of neck scale and slicker plate provers and 2) the volume transfer method for neck scale prover to neck scale prover.

It is important to note that the uncertainty analyses given herein are for the volume determinations made by the NIST volume calibration service and do not include components that must be considered by the customer during subsequent use of the prover as a working standard for volume. Some of these sources of uncertainty are: errors in reading the neck scale, viscosity effects caused by temperature or fluid changes, the effects of dirty surfaces in the prover or in the sight glass that cause changes in the draining or changes in the behavior of the meniscus, dents in the prover, poor leveling of the prover, leaks, incorrect relationship between neck scale increments and neck diameter.

The relative uncertainty of the volume measurement process may be expressed as [ISO 1996 and Coleman and Steele, 1999],

$$U_V = k\left[\sum_{i=1}^{n}\left(\frac{1}{V}\frac{\partial V}{\partial x_i}u(x_i)\right)^2 + \frac{2}{V^2}\sum_{i=1}^{n}\sum_{j=i+1}^{n-1}\frac{\partial V}{\partial x_i}\frac{\partial V}{\partial x_j}u(x_i)u(x_j)r(x_i,x_j) + \frac{\sigma_V^2}{V^2}\right]^{\frac{1}{2}} \quad (19)$$

where:

- V represents any of the data reduction equations used to calculate the prover volume,
- x_i are the variables in V,
- $u(x_i)$ are the uncertainties in the variables in V,
- i and j are indices that are never equal
- n is the number of variables in V,
- σ_V^2 is the variance of replicated determinations of prover volume, and
- k is a coverage factor, typically about 2, selected to assure a 95 % confidence interval level for the measurement process and consistent with NIST policy [Taylor and Kuyatt, 1994],
- $r(x_i,x_j)$ is the correlation coefficient between variables x_i and x_j.

The second term on the right hand side of equation 19 is the correlation term used in the estimate of the uncertainty due to bias when any of the terms in a data reduction equation are correlated. For example, correlation can occur when:

(a) one instrument is used to measure more than one term in an equation for V;
(b) the same expression is used to calculate more than one term in an equation for V; and
(c) more than one instrument used to measure terms in an equation for V are calibrated by the same standard.

For the gravimetric methods of volume calibration, the two indicated masses are measured with the same weigh scale and the mass uncertainties have a high degree of correlation. The two masses are subtracted from one another and taking the correlation of their uncertainty into account would have the effect of reducing the combined uncertainty of the volume determination. In the following uncertainty analysis, the mass uncertainties were treated as uncorrelated, resulting in the most conservative estimate of the combined uncertainty.

When the same working standard volume is used repeatedly to fill a larger prover under test via the volume transfer method, the uncertainty of $V_{\text{deliver}}(T_w)$ for the working standard volume is correlated. Because the volumes are added, considering their uncertainty fully correlated is a more conservative approach and that is the approach we will follow in the following uncertainty analysis.

For the volume transfer method, the correlated temperature uncertainties are negligible relative to the uncorrelated ones, and therefore they are treated as completely uncorrelated (most conservative treatment).

7.1 Uncertainty of the Gravimetric Method

Because equation 2 is implicit in these equations for V and its differentiation over the five temperature dependant terms would be unwieldy, its first order approximation is used in this uncertainty analysis:

$$\rho(T_w) = \rho_0 [1 - \beta_w (T_w - T_0)] \quad (20)$$

where

ρ_0 is the density of the distilled water at the water reference temperature, T_0 of 3.9818 °C, and

β_w is the volumetric thermal expansion coefficient for water.

While the accuracy of equation 20 is inferior to that of equation 2, it is satisfactory for uncertainty analysis.

The complete data reduction equation for the gravimetric determination of $V_{contain}(T_{ref})$ is obtained by combining equations 3, 5, and 20. The result is

$$V_{contain}(T_{ref}) = \left[\frac{(m_f - m_e)}{\rho_0 [1 - \beta_w (T_w - T_0)] - \rho_{air}} - h \right] [1 - \beta_t (T_w - T_{ref})]. \quad (21)$$

The partial derivatives for equation 21 and their values are listed in Tables 1 through 4 for four volume provers to show example uncertainty analyses. Uncertainties for a customer calibration may be different from those given here, especially those related to the water height in a volume prover with a neck scale. Defined values having zero uncertainty are T_{ref} and T_0.

The Guide to the Expression of Uncertainty in Measurement [ISO 1996] suggests classifying the uncertainty components as: <u>Type A</u>, uncertainties which are evaluated by statistical methods, and <u>Type B</u>, uncertainties which are evaluated by other means. Once a value for the uncertainty for each component is determined, it receives the identical mathematical treatment in equation 19 whether it is either Type A or B. The sources of uncertainties, and their types, for equation 21 are:

- $u(h)$ is one fourth of the smallest division of the neck scale (read with a magnifier fitted with an anti-parallax device) (Type B),
- $u(m_f)$, $u(m_e)$ are based on analysis of calibration records and uncertainty of the calibration process for the weigh scales, normally less than 5 g for the 600 kg weigh scale, and less than 1 g for the 60 kg scale (Type B).
- $u(\rho_0)$, $u(\beta_w)$ are based on the water density equation [Patterson and Morris, 1994] (Type B),
- $u(\beta_t)$ was provided by the NBS Crystallography Section (Type B),
- $u(T_w)$ is based on calibration records, the uncertainty of the temperature calibration process within the Fluid Metrology Group, and the normal variation of temperature within the volume prover (Type B),
- $u(\rho_{air})$ is from the uncertainty of the equation for air density [Jaeger and Davis, 1984] and the uncertainties in the measurement of room temperature, pressure, and relative humidity (Type B),

$u(v)$ is the uncertainty due to viscosity effects on delivered volume (Type B), and

σ_V^2 is the variance of replicated determinations of prover volume (Type A).

The expression for delivered volume has an additional uncertainty (that does not apply to the contained volume) due to the effects of temperature on water viscosity and the difference in completeness of draining of the prover at T_w and T_{ref}. The uncertainty in the delivered volume due to viscosity effects is assumed to have a rectangular distribution with $\pm a = V_v$, so that the standard uncertainty due to the viscosity correction is $V_v/(\sqrt{3})$ [Taylor and Kuyatt, 1994].

The coverage factor, k, in equation 18 is selected to assure a 95 % confidence interval level for the measurement process, consistent with NIST policy [Taylor and Kuyatt, 1994]. Small sample (< 10) statistics states that for a combined standard uncertainty, U_V, the "effective degrees of freedom" v_{eff} of U_V, which is approximated by appropriately combining the degrees of freedom of its components, should be used to establish the value of the coverage factor. Taylor and Kuyatt [1994] recommend that the Welch-Satterthwaite formula be used to determine the effective degrees of freedom for the measurement process. Based on the value obtained, the appropriate coverage factor, k, can be obtained from the t-distribution for degrees of freedom for a Gaussian distribution with a confidence interval of 95 %.

Liquid Volume Calibration Service NIST SP 250-72

Tables 1 through 4 show data reduction spreadsheets and uncertainty calculations for four volume provers: 42 gal, 21 gal, 5 gal (all with neck scales), and a 5 gal slicker plate volume prover. Explanatory notes for Table 1 are given below.

A) Volume prover identification, size, volume per smallest scale division, provided by manufacturer.
B) Weigh scale calibration coefficients.
C) Temperature sensor calibration coefficients.
D) Two water temperatures and their average, used to calculate water density.
E) Two air temperatures and their average, the barometric pressure, and relative humidity, used to calculate air density for buoyancy corrections to mass.
F) Weights of the volume prover dry (one measurement entered 5 times), full (5 repeats), and drained (5 repeats).
G) The neck scale readings for the 5 full weight measurements.
H) Calculated prover volumes for the 5 repeats at room temperature, adjusted to scale reading of zero, without a viscosity correction.
I) Prover volumes for the 5 repeats, corrected for thermal expansion to the reference temperature, in various units, contained and delivered.
J) Viscosity values at the average calibration temperature and reference temperature (for the viscosity correction).
K) Averages of the 5 repeats, contained and delivered volumes (without and with viscosity correction).
L) Repeatability of the 5 volume measurements (a Type A component of the calibration uncertainty) and the expanded uncertainty.
M) Coverage factor and degrees of freedom from the Welch-Satterthwaite analysis.
N) Values of variables used in the volume calculation needed for the uncertainty analysis.
O) Analytical expressions and values for the sensitivity coefficients, and the magnitude of the uncertainty for each uncertainty component.
P) Uncertainty contribution (in parts in 10^6) for each uncertainty component.
Q) Neck scale readings, correct neck scale divisions (calculated based on delivered volume calibration), and differences between them (corrections), for five liquid levels in the neck scale.
R) Slope and intercept of a best fit line to correct neck scale readings to calculated delivered volume (in various units).
S) A plot of neck scale correction versus neck scale reading (in neck scale divisions), both the measured points and the best fit line.

Table 1. Sample spreadsheet with uncertainty analysis for the gravimetric determination of $V_{\text{deliver}}(T_{\text{ref}})$ of a 42 gal (160 L) graduated neck scale volume prover. See text for key.

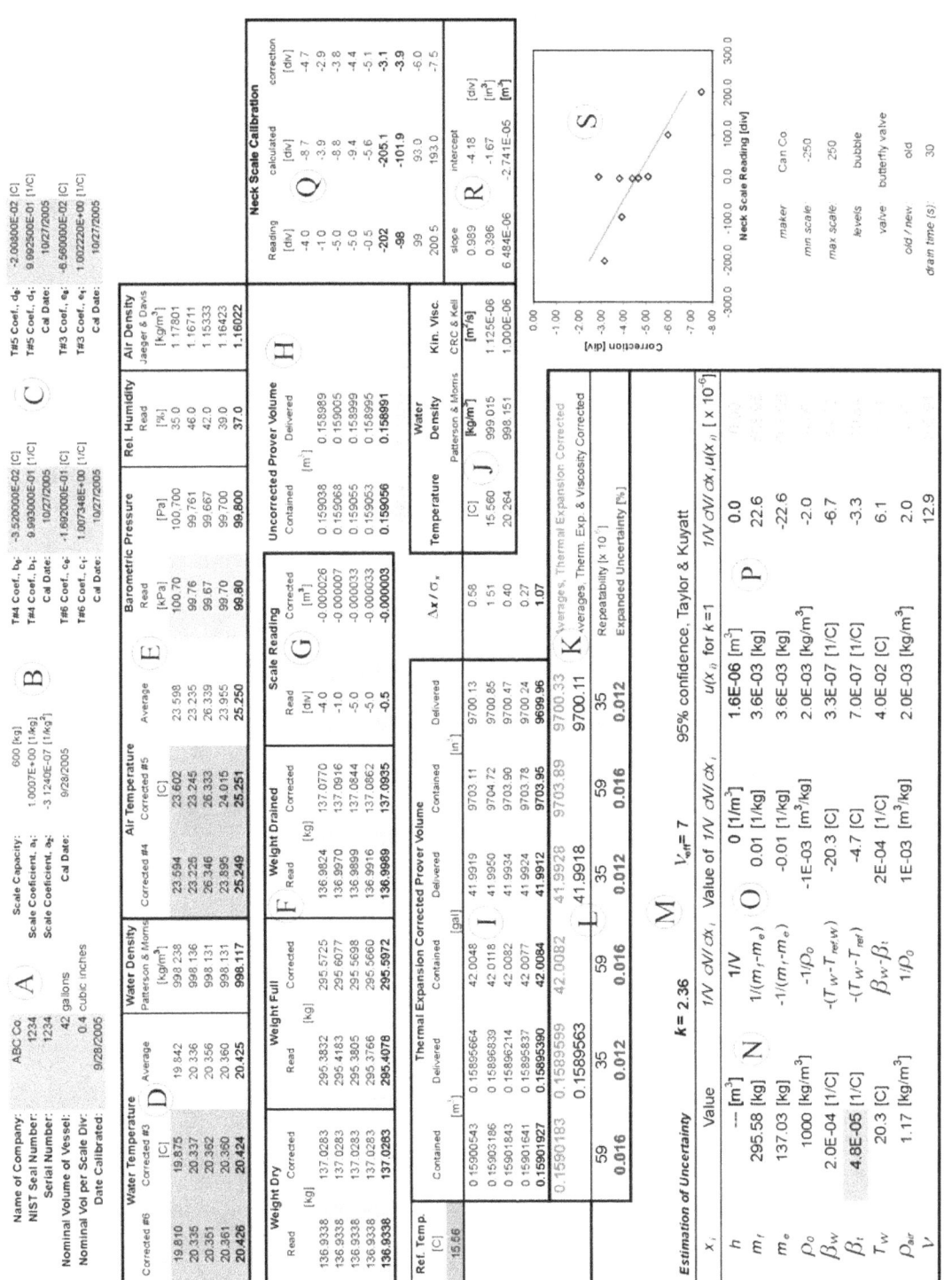

Table 2. Sample spreadsheet with uncertainty analysis for the gravimetric determination of $V_{\text{deliver}}(T_{\text{ref}})$ of a 21 gal (80 L) graduated neck scale volume prover.

Liquid Volume Calibration Service NIST SP 250-72

Table 3. Sample spreadsheet with uncertainty analysis for the gravimetric determination of $V_{\text{deliver}}(T_{\text{ref}})$ of a 5 gal (19 L) graduated neck scale volume prover.

Table 4. Sample spreadsheet with uncertainty analysis for the gravimetric determination of $V_{\text{deliver}}(T_{\text{ref}})$ of a 5 gal (19 L) slicker plate volume prover.

7.2 Uncertainty of the Volume Transfer Method

In this section, we will analyze the uncertainty of the volume transfer method using the example of the calibration of a 450 gal (1700 L) prover with a 100 gal (378 L) working standard prover (transferred 4 times) and a 50 gal (189 L) working standard (transferred once). The basis equation for the uncertainty analysis is equation 10' with the water density approximation of equation 20.

The component uncertainties, and their types, are:

$u(h_t)$, $u(h_A)$, $u(h_B)$	are one fourth of the smallest division of the neck scale of the prover under test and of the two working standard provers (Type B),
$u(V_A)$, $u(V_B)$	are based on the calibration records and uncertainty of the calibration process for the working standard volume provers (Type B),
$u(\rho_t)$, $u(\rho_A)$, $u(\rho_B)$, $u(\beta_w)$	are from the water density equation [Patterson and Morris, 1994] and uncertainties in the measurement of temperature (Type B),
$u(T_t)$, $u(T_A)$, $u(T_B)$	are based on calibration records, the uncertainty of the temperature calibration process within the Fluid Metrology Group, and the normal variation of temperature within the volume prover (Type B),
$u(\beta_t)$	was provided by the NBS Crystallography Section (Type B),
$u(\rho_0)$, $u(\beta_w)$	are based on the water density equation [Patterson and Morris, 1994] (Type B),
$u(v)$	is the uncertainty due to viscosity effects on delivered volume (Type B), and
σ_V^2	is the variance of replicated determinations of prover volume (Type A).

Tables 5 and 6 show sample calculations for the volume transfer method. Table 5 shows the summation of working standard volumes for one of the five repeats. Table 6 summarizes the results for all five repeats, including averaging, statistical calculations, and the uncertainty calculation. Five values for the contained volume are calculated even though only one volume transfer was done with the prover dry: the difference between contained and delivered volume from an initial, dry run is added to the subsequent delivered volume to obtain the final four contained volumes.

Differentiation of equation 10' and elimination of terms that are small compared to unity leads to the normalized sensitivity coefficients listed in Table 6. The coefficients of thermal expansion for all three provers were assumed to be equal (β_t) in this uncertainty analysis. For the example in Table 6, a simplified uncertainty analysis that considered only the components due to V_A, V_B, h_t, and the test repeatability would give a good approximation (0.033 % versus 0.035 %). However, in cases with larger temperature uncertainty, the temperature components will dominate and they should not be ignored.

Explanatory notes for Tables 5 and 6 are given below.

A) Volume prover identification, size, volume per smallest scale division.
B) Temperature sensor calibration coefficients.
C) Water temperatures for each of the volume working standards before transfer to volume prover under test, used to calculate water density. Last row is the temperature of the prover under test when full.
D) Water densities for working standards and prover under test.
E) Volume, reference temperature, etc. for the working standard provers.
F) Neck scale readings for the working standards and the prover under test.
G) Volumes and masses for the working standards and the total transferred to the prover under test.
H) A single volume determination result for the prover under test at test temperature and at the reference temperature.
I) Results for five individual volume determinations for the prover under test.
J) The average for the repeated measurements, their standard deviation, and the uncertainty of the calibration (based on calculations shown below).
K) Coverage factor and effective degrees of freedom.
L) Values of variables used for the uncertainty analysis.
M) Analytical expressions and values for the sensitivity coefficients, and the magnitude of the uncertainty for each uncertainty component.
N) Uncertainty contribution (in parts in 10^6) for each uncertainty component.

Table 5. Sample of the worksheet that totalizes the mass for the volume transfer method, 4 × 100 gal + 1 × 50 gal ≈ 450 gal. See text for key.

Name of Company:	ABC Co.	
NIST Seal Number:	1234	
Serial Number:	1234	A
Nominal Volume of Vessel:	1.7 [m^3]	
Nominal Vol per Scale Div:	2.50E-04 [m^3]	
Date Calibrated:	7/11/2004	
T#6 Coef., c_0:	0.000E+00 [C]	
T#6 Coef., c_1:	1.000E+00 [1/C]	B
Cal. Date:	5/10/2004	

Water Temperature

Read @ #6 [°C]	Corrected Ave.
25.600	
25.650	C
25.440	
25.550	
25.990	
25.610	

Water Density

Patterson & Morris [kg/m^3]	
996.890	
996.877	
996.932	D
996.903	
996.787	
996.888	Final

Standard Test Measures

NIST Prop. No.	Cal. Date	Ref. Volume [m^3]	Ref. Temp. [°C]	Therm. Exp. Coef. [1/°C]	Uncertainty, k=1 [±m^3]	Volume per Div. [in^3/div]
NBS 152550	6/23/2004	0.37850714	15.56	4.8E-05	0.00003295	2.000
NBS 152550	6/23/2004	0.37850714	15.56	4.8E-05	0.00003295	2.000
NBS 152550	6/23/2004	0.37850714	15.56	4.8E-05	0.00003295	2.000
NBS 62069	6/28/2004	0.18919155	15.56	4.8E-05	0.00002660	1.000

E (Ref. Volume column label region)

Scale Reading

Read [div]	Corrected [m^3]
-57.0	-0.00186812
-57.2	-0.00187468
-1.2	-0.00003933
-1.8	-0.00005899
5.0	0.00008194
-4.0	-0.00100000

F

Water Delivered

Volume [m^3]	Mass [kg]
0.37663902	375.6476
0.37663246	375.6370
0.37846781	377.4845
0.37844815	377.4560
0.18927349	188.7593
Final	1694.9845

G

Water Ref. Properties

Temperature [°C]	Density Patterson & Morris [kg/m^3]
15.56	999.015

Vessel

Therm. Exp. Coef. [1/°C]
4.8E-05

Vessel Volume @ Temp.

Delivered [m^3]	Delivered [gal]	Delivered [L]
1.70128	449.430	1701.28

H

Vessel Volume @ Ref. Temp.

Delivered [m^3]	Delivered [gal]	Delivered [L]
1.70046	449.214	1700.46

Table 6. Summary of the five repeated measurements, averages, and sample uncertainty analysis for the volume transfer method. See text for key.

Name of Company:	ABC Co.	A
NIST Seal Number:	1234	
Serial Number:	1234	
Nominal Volume of Vessel:	1.700 [m³]	
Nominal Vol per Scale Div:	2.50E-04 [m³]	
Date Calibrated:	7/11/2005	

# of Tranfers	
N_A	4
N_B	1

Ref. Temp. [C]
15.56

Thermal Expansion Corrected Prover Volume

Contained [m³]	Delivered	Contained [gal]	Delivered	Contained [L]	Delivered	D_x / s_x	Temperature [C]	Water Density Patterson & Morris [kg/m³]	Kin. Visc. CRC & Kell [m²/s]
1.70086	1.70046	449.214	449.214	1700.86	1700.46	1.45	15.560	999.015	1.125E-06
1.70114	1.70074	449.289	449.289	1701.14	1700.74	0.08	25.680	996.869	8.829E-07
1.70110	1.70070	449.277	449.277	1701.10	1700.70	0.30			
1.70130	1.70090	449.330	449.330	1701.30	1700.90	0.67			
1.70140	1.70100	449.357	449.357	1701.40	1701.00	1.16			

I

1.70116	1.70076	449.2936	449.294	1701.16	1700.76	Averages, Thermal Expansion Corrected
	1.70071		449.280		1700.71	Averages, Therm. Exp. & Viscosity Corrected
121	121	121	121	121	121	Repeatability [x 10⁻⁶]
0.035	0.035	0.035	0.035	0.035	0.035	Expanded Uncertainty [%]

J

Estimation of Uncertainty k= 2.18 V_{eff}= 12   95% confidence, Taylor & Kuyatt

x_i	Value	$1/V \cdot \partial V/\partial x_i$	Value	dx_i for k=1	$1/V \cdot \partial V/\partial x_i \cdot dx_i$ [x 10⁶]	N
h_1	0.00 [m³]	$1/V_t$	5.88E-01 [1/m³]	6.25E-05 [m³]	37	
h_B	0.00 [m³]	N_B/V_t	5.88E-01 [1/m³]	1.64E-06 [m³]	1	
h_A	0.00 [m³]	N_A/V_t	2.35E+00 [1/m³]	3.28E-06 [m³]	8	
V_B	0.1893 [m³]	N_B/V_t	5.88E-01 [1/m³]	2.66E-05 [m³]	16	
V_A	0.3766 [m³]	N_A/V_t	2.35E+00 [1/m³]	3.29E-05 [m³]	78	
T_t	25.68 [°C]	$(\beta_t+\beta_w)N_t$	1.46E-04 [1/°C]	2.0E-01 [C]	29	
T_B	25.91 [°C]	$[N_B V_B(\beta+\beta)]/V_t$	2.76E-05 [1/°C]	2.0E-01 [C]	6	
T_A	25.70 [°C]	$[N_A V_A(\beta+\beta)]/V_t$	2.20E-04 [1/°C]	2.0E-01 [C]	44	
β_t (prover)	4.8E-05 [1/C]	$(N_A V_A(-T_1+T_{ref}+T_A-T_{CA})+N_B V_B(-T_1+T_{ref}+T_B-T_{CB}))/V_t$	8.10E+00 [°C]	7.0E-07 [1/C]	6	
β_w (water)	2.0E-04 [1/C]	$(N_A V_A(T_A-T_1)+N_B V_B(T_B-T_1))/V_t$	3.89E-02 [°C]	2.0E-06 [1/C]	0	
ρ_0	1000 [kg/m³]	$-1/\rho_0$	1E-03 [m³/kg]	2.0E-03 [kg/m³]	2	
v					-17	

K   L   M

8. References

Bird, R. B., Stewart, W. E., and Lightfoot E. N., *Transport Phenomena*, Wiley, New York, 1960, pp. 69.

Chemical Rubber Publishing Company, *Handbook of Chemistry and Physics*, 40th Edition, Cleveland, 1959, pp. 2156.

Coleman, H. W. and Steele, W. G., *Experimentation and Uncertainty Analysis for Engineers*, (John Wiley and Sons, New York, 2nd edition, 1999).

Harris, G. L., *Specifications and Tolerances for Reference Standards and Field Standard Weights and Measures: 3. Specifications and Tolerances for Graduated Neck Type Volumetric Field Standards*, NIST Handbook 105-3, National Institute of Standards and Technology, Gaithersburg, MD, June, 1997.

Houser, J. F., *Procedures for the Calibration of Volumetric Test Measures*, NBSIR 73-287, National Bureau of Standards, Washington, D.C. (1973).

International Organization for Standardization, *Guide to the Expression of Uncertainty in Measurement*, Switzerland, 1996.

Jaeger, K. B., and Davis, R. S., *A Primer for Mass Metrology*, National Bureau of Standards (U.S.), NBS Special Publication 700-1, Industrial Measurement Series, 1984, 79 p.

Kell, G. S., *Density, Thermal Expansively, and Compressibility of Liquid Water from 0°C to 150°C: Correlations and Tables for Atmospheric Pressure and Saturation Reviewed and Expressed on 1968 Temperature Scale*, J. Chem. Eng. Data, 20, 97-105, (1975).

Marshall, J. L., *NIST Calibration Services Users Guide*, NIST Special Publication 250, National Institute of Standards and Technology, Gaithersburg, MD, January, 1998.

Patterson J. B., and Morris E. C., *Measurement of Absolute Water Density, 1°C to 40°C*, Metrologia, 31, 277-288, (1994).

Taylor, B. N., and Kuyatt, C. E., *Guidelines for Evaluating and Expressing the Uncertainty of NIST Measurement Results*, NIST Technical Note 1297, (U.S. Government Printing Office, Washington D.C., 1994).

Appendix A. Calibration of Weigh Scales

The traceability of liquid volume measurements to the national standards of mass is through a 65 kg set of weights calibrated by the NIST Acoustics, Mass, and Vibrations Group. Weights from the 65 kg set are used to calibrate the 60 kg weigh scale. Weights from the 65 kg set and the 60 kg weigh scale are used to calibrate a set of 45 kg weights, which in turn, are used to calibrate the 600 kg weigh scale. The process for calibrating the 45 kg weights is given below. Following that are the steps necessary for calibrating the 60 kg and 600 kg weigh scales with the reference masses.

Calibration of the 45 kg Reference Masses

1. Measure the ambient air temperature, the barometric pressure, and the relative humidity. Calculate the air density, ρ_{air}, using equation 1.

2. Set the 60 kg weigh scale to zero.

3. Load 45 kg mass, m_{std}, from the 65 kg weight set on the weigh scale platform and record the reading as O_1. The reading of the weigh scale is the difference of the downward force due to gravity and the upward force due to air buoyancy.

$$O_1 g = m_{std} g - V_{std} \rho_{air} g = m_{std}\left(1 - \frac{\rho_{air}}{\rho_{std}}\right) g \tag{1A}$$

where:
   g is the local acceleration due to gravity,
   V_{std} is the volume of m_{std}, and
   ρ_{std} is the density of the metal from which m_{std} is made.

Since g appears in all terms, we have, by dividing by g

$$O_1 = m_{std}\left(1 - \frac{\rho_{air}}{\rho_{std}}\right) \tag{2A}$$

The double substitution weighing design requires three more weighings and three more equations, analogous to equation 1A, to describe them. Since g is a common factor in all terms of all three equations, the equations are written the form analogous to equation 2A where division by g is understood.

4. Remove m_{std}, reset the weigh scale to zero, load a 45 kg weight to be calibrated, m_{45}, and record the reading as O_2,

$$O_2 = m_{45}\left(1 - \frac{\rho_{air}}{\rho_{45}}\right)$$

where ρ_{45} is the density of the metal from which the 45 kg weights are made,

5. Add a 0.2 kg sensitivity weight, m_s, and record the reading as O_3,

$$O_3 = O_2 + m_s\left(1 - \frac{\rho_{air}}{\rho_s}\right)$$

where ρ_s is the density of the metal from which the sensitivity weight is made,

6. Remove m_{45}, reset the weigh scale to zero, load m_{std}, and record the reading for m_{std} and m_s as O_4,

$$O_4 = O_1 + m_s\left(1 - \frac{\rho_{air}}{\rho_s}\right)$$

By adding and subtracting the last four equations, we obtain

$$\frac{(O_2 + O_3 - O_1 - O_4)}{2} \times \frac{m_s\left(1 - \frac{\rho_{air}}{\rho_s}\right)}{(O_3 - O_2)} = m_{45}\left(1 - \frac{\rho_{air}}{\rho_{45}}\right) - m_{std}\left(1 - \frac{\rho_{air}}{\rho_{std}}\right)$$

where the term on the left hand side has been multiplied by 1 in order to include the measurements involving the sensitivity weight, i.e.,

$$1 = \frac{m_s\left(1 - \frac{\rho_{air}}{\rho_s}\right)}{(O_3 - O_2)}$$

The calibration equation for the 45 kg weights is obtained by solving for m_{45},

$$m_{45} = \frac{\dfrac{(O_2 + O_3 - O_1 - O_4)}{2(O_3 - O_2)} m_s\left(1 - \dfrac{\rho_{air}}{\rho_s}\right) + m_{std}\left(1 - \dfrac{\rho_{air}}{\rho_{std}}\right)}{\left(1 - \dfrac{\rho_{air}}{\rho_{45}}\right)}$$

Calibration of the 60 kg and 600 kg Weigh Scales

The 60 kg weigh scale is calibrated by placing weights from the 65 kg set on the platform in 5 kg increments for the up-loading cycle. The down-loading increments are done by removing 10 kg and restoring 5 kg per step in order to nullify the effects of hysteresis.

The 600 kg weigh scale is calibrated by placing 45 kg weights on the platform, one weight per increment for the up-loading cycle. The down-loading cycle is done by removing 90 kg and restoring 45 kg per increment in order to nullify the effects of hysteresis.

The calibration process for a weigh scale generates a series of weigh scale readings, O, for a series of calibrated mass values, m. The calibration equation for either weigh scale is obtained by fitting the mass values, adjusted for air buoyancy, as a function of the weigh scale readings,

$$m\left(1 - \frac{\rho_{air}}{\rho_m}\right) = A + BO + CO^2 + \ldots$$

Thus the indicated mass, m_{ind}, (not air buoyancy adjusted) for an object being weighed on a calibrated weigh scale is

$$m_{ind} = A + BO + CO^2 + \ldots \tag{3A}$$

The step-by-step procedure is:

1. Measure and record the initial room air temperature, atmospheric pressure, and relative humidity.

2. Place the first reference mass on the weigh scale near the center of the weigh scale pan. (The order in which the reference masses are applied to the scale is indicated on the data collection worksheet.) Wait approximately 10 seconds for stability of the mass indicated by the weigh scale and record the indicated mass.

3. Add the subsequent reference masses, one by one, near the center of the weigh scale, recording the readings of the weigh scale at each step.

4. Measure and record the final room air temperature, atmospheric pressure, and relative humidity.

5. Process the calibration data. Apply buoyancy corrections (using equation 1 for air density) to calculate the apparent masses for the reference masses. Perform a 2nd order least squares fit to obtain a polynomial that calibrates the weigh scale readings. This polynomial is applied to the readings of the weigh scale gathered during calibration of the prover volume.

Appendix B. Derivation of Equations for Gravimetric Method for Neck Scale Provers

In the gravimetric procedures for neck scale volume provers listed in Section 4.1.1, the weigh scales readings for the dry volume prover, O_e, the filled volume prover, O_f, and the drained volume prover, O_{ed}, are recorded in steps 6, 11, and 14, respectively. These readings are converted to indicated masses m_e, m_f, and m_d in steps 16-18 via equation 3A.

The indicated mass for the dry measure is

$$m_e = M_e \left(1 - \frac{\rho_{air}}{\rho_e}\right) \tag{1B}$$

where:

M_e is the mass of the dry measure,

ρ_{air} is the density of the air, and

ρ_e is the density of the dry measure.

For the measure filled with water up to the zero reference mark, the indicated mass is

$$m_f = M_e \left(1 - \frac{\rho_{air}}{\rho_e}\right) + M_{w,contain}\left(1 - \frac{\rho_{air}}{\rho(T_w)}\right) = m_e + M_{w,contain}\left(1 - \frac{\rho_{air}}{\rho(T_w)}\right) \tag{2B}$$

where :

$M_{w,contain}$ is the mass of the contained water and

$\rho(T_w)$ is the density of the water at temperature T_w.

Solving equation 2B for $M_{w,contain}$,

$$M_{w,contain} = \frac{(m_f - m_e)\rho(T_w)}{\rho(T_w) - \rho_{air}} \tag{3B}$$

From the definition of density,

$$V_{contain}(T_w) = \frac{M_{w,contain}}{\rho(T_w)} = \frac{m_f - m_e}{\rho(T_w) - \rho_{air}} - h \tag{4B}$$

where h is the deviation of the meniscus from the zero mark on the neck scale, in volume units.

The indicated mass for the drained measure is

$$m_d = m_e + M_{w,residual}\left(1 - \frac{\rho_{air}}{\rho(T_w)}\right) \tag{5B}$$

where $M_{w,residual}$ is the water remaining in the measure after the valve is closed at the end of the specified drain time.

For the measure filled with water up to the zero reference mark,

$$m_f = m_e + M_{w,residual}\left(1 - \frac{\rho_{air}}{\rho(T_w)}\right) + M_{w,deliver}\left(1 - \frac{\rho_{air}}{\rho(T_w)}\right) = m_d + M_{w,deliver}\left(1 - \frac{\rho_{air}}{\rho(T_w)}\right) \quad (6B)$$

Solving for $M_{w,deliver}$ and combining with the definition of density, we obtain

$$V_{deliver}(T_w) = \frac{m_f - m_d}{\rho(T_w) - \rho_{air}} - h. \quad (7B)$$

Appendix C. Derivation of Equation for the Volume Transfer Method

The transfer method of volume calibration is accomplished by emptying a known volume of water from a working standard measure into the measure to be calibrated. For the present purpose, it is assumed that:

1. The working standard is a neck-scale measure.
2. The delivered volume of the working standard, V_s, has been determined by the gravimetric method at $T_{w,c}$ via equation 5.
3. The measure under test is a neck-scale measure of the same nominal capacity as the working standard so that only one emptying of the working standard is required.

The volume of the water in the working standard is

$$(V_s + h_s)[1 + \beta_s(T_{w,s} - T_{ref,s})] = \frac{m}{\rho(T_{w,s})} \tag{1C}$$

where:

V_s is the volume of the working standard filled up to the proximity of the zero reference mark,

h_s is the deviation from zero of the neck scale reading of the working standard, in volume units, with the appropriate algebraic sign applied,

β_s is the volumetric thermal expansion coefficient of the working standard,

$T_{w,s}$ is the temperature of the water in the working standard,

$T_{ref,s}$ is the temperature at which the working standard was calibrated,

m is the mass of the water transferred, and

$\rho(T_{w,s})$ is the density of the water in the working standard.

After the transfer, the volume of the water in the measure under test, V_t, is

$$(V_t + h_t) = \frac{m}{\rho_{w,t}} \tag{2C}$$

where:

h_t is the neck scale reading of the measure under test, in volume units, with the appropriate algebraic sign applied,

$T_{w,t}$ is the temperature of the water in the measure under test, and

$\rho(T_{w,t})$ is the density of the water in the working standard.

Since mass is conserved in the transfer process, equations 1C and 2C are solved for m and the resulting expressions are equated and solved for V_t.

$$V_t(T_{w,t}) = \frac{\rho(T_{w,s})(V_s + h_s)[1 + \beta_s(T_{w,s} - T_{ref,s})]}{\rho(T_{w,t})} - h_t \ . \tag{3C}$$

This expression will yield contained volume at $T_{w,t}$ if the inside walls of the measure under test were dry at the outset of the calibration, or delivered volume at $T_{w,t}$ if the inside walls of the prover were pre-wetted at the outset of the calibration.

National Institute of Standards and Technology
Technology Administration, U.S. Department of Commerce

REPORT OF CALIBRATION

FOR

A ONE HUNDRED (100) GALLON VESSEL
(Graduated Neck Type)

April 28, 2005

Manufacturer: Test Measure Co. NIST Seal Number: 1234
Gaithersburg, MD Material: Stainless Steel
Maker Number: ABC123

submitted by

Test Measure Co.
123 West St.
Gaithersburg, MD 20877

(Reference: Purchase Order Number ABC123, April 1, 2005)

The internal volume of the vessel described above has been determined by the gravimetric method [1]. The gravimetric method requires weighing the vessel dry and empty and re-weighing it when filled with a fluid of known density. The internal or contained volume was determined in this way and the value is given in Table 1 using the requested units. The fluid used was water and the vessel was leveled before determining the volume.

To determine the delivered volume, the contained volume is poured from the vessel by opening the valve at the bottom of the vessel. When this flow finishes, the valve is held open for 30 seconds to complete the drain procedure. Subsequent re-weighing completes the gravimetric procedure and enables calculation of the delivered volume, also given in Table 1. Both the contained and delivered volumes are given for the scale reading of zero (0) and have been corrected for the reference temperature of 15.56 °C (60 °F) assuming a volumetric coefficient of expansion of 0.0000477 per °C (0.0000265 per °F).

---

[1] Bean, V. E., Espina, P. I., Wright, J. D., Houser, J. F., Sheckels, S. D., and Johnson, A. N., "NIST Calibration Services for Liquid Volume", NIST Special Publication 250-72, National Institute of Standards and Technology, November 25, 2009.

NIST Test Number: 836/123456-05-01
Calibration Date: April 26, 2005 by Sherry Sheckels and John Wright

REPORT OF CALIBRATION
Test Measure Co.

NIST Seal No.: 1234
Purchase Order No. ABC123

Table 1. Contained and delivered volumes for the tested vessel for a scale reading [2] of zero.

	Volume Contained	*Volume Delivered*
gal at 60 °F	99.9924	**99.9656**
in^3 at 60 °F	23098.25	**23092.07**

The expanded uncertainty in the measured volume is ±0.011 % and the repeatability of the 5 measurements was 39 parts in 10^6. The uncertainty was calculated according to References [1] and [3] with a 95 % confidence level [4] and is traceable to NIST mass, temperature, pressure, and humidity standards, and a NIST water density determination.

Below are pictures of the identification plate of the calibrated vessel and its drain valve. Enclosed at the end of this report is a spreadsheet outlining all the measurements made during this calibration and the computed results.

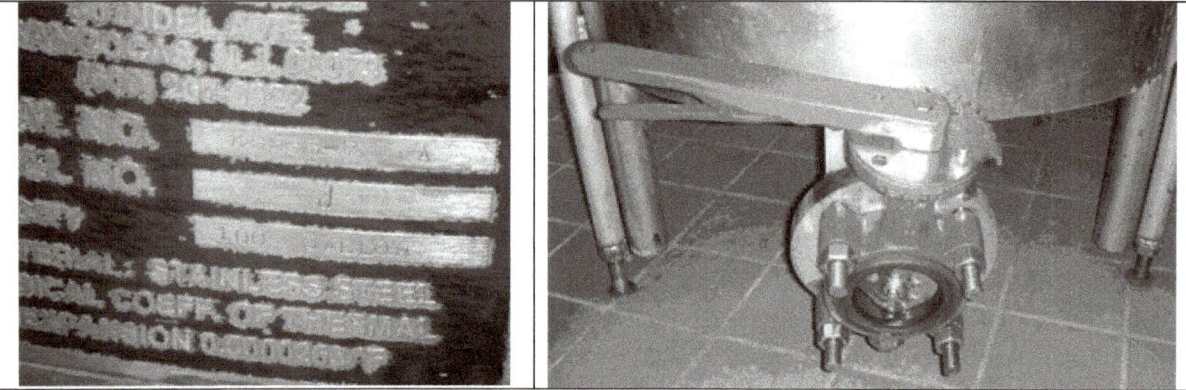

Figure 1. Photographs of the identification plate of the calibrated prover and its drain valve.

---

[2] The scale reading is determined by the intersection of the horizontal plane, tangent to the bottom of the meniscus on the gauge tube. For this prover, the neck scale range was from -225 to +200 and each division is labeled by the manufacturer to be equivalent to 5 in^3.

[3] Taylor, B. N., and Kuyatt, C. E., "Guidelines for Evaluating and Expressing Uncertainty of NIST Measurement Results," NIST Technical Note 1297, National Institute of Standards and Technology, January, 1994.

[4] Coverage factor, *k* of 2.26 for 9 effective degrees of freedom.

NIST Test Number: 836/123456-05-01
Calibration Date: April 26, 2005 by Sherry Sheckels and John Wright

REPORT OF CALIBRATION
Test Measure Co.

NIST Seal No.: 1234
Purchase Order No. ABC123

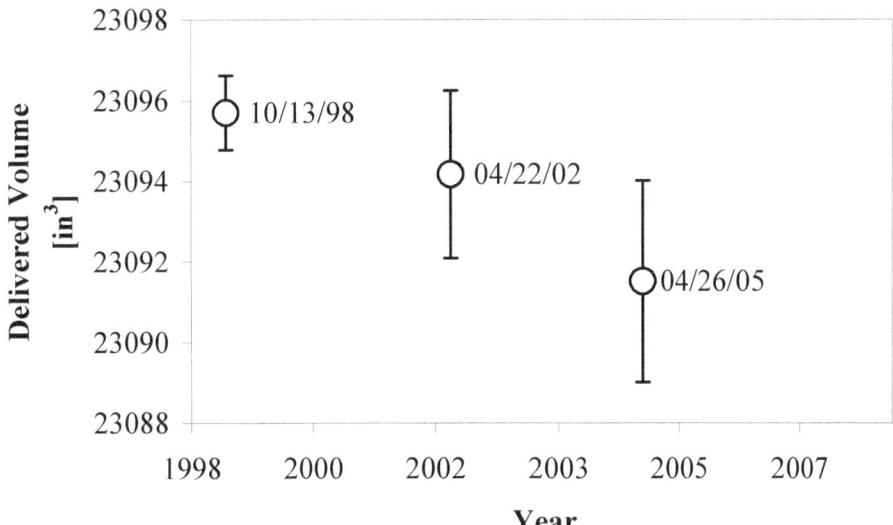

Figure 2. Calibration control chart for 100 gal graduated neck volume prover SN ABC123.

Table 2. Results of prior calibrations for the delivered volume.

Date	Delivered Volume [in^3]	Difference from Prior [in^3]	Degree of Equivalence [-]
04/26/05	23092.07	-2.10	0.36
04/22/02	23094.17	-1.53	0.26
10/13/98	23095.70		

For the Director,
National Institute of Standards and Technology

Dr. John D. Wright
Project Leader, Fluid Metrology Group
Process Measurements Division
Chemical Science and Technology Laboratory
National Institute of Standards and Technology

Sherry Sheckels
Calibration Technician, Fluid Metrology Group
Process Measurements Division
Chemical Science and Technology Laboratory
National Institute of Standards and Technology

NIST Test Number: 836/123456-05-01
Calibration Date: April 26, 2005 by Sherry Sheckels and John Wright

REPORT OF CALIBRATION
Test Measure Co.

NIST Seal No.: 1234
Purchase Order No. ABC123

Name of Company: Test Measure Co.
NIST Seal Number: 1234
Serial Number: ABC123
Volume of Vessel: 0.3785412 [m³]
Volume per Scale Division: 5.000 [in³/div]
Date Calibrated: 4/26/2005

Scale Capacity: 600 [kg]
Scale Coefficient, a_1: 1.0008E+00 [1/kg]
Scale Coefficient, a_2: -3.2489E-07 [1/kg²]
Cal Date: 4/21/2005

T#4 Coef., b_0:	3.330000E-02 [C]
T#4 Coef., b_1:	1.000755E+00 [1/C]
Cal Date:	10/7/2004
T#6 Coef., c_0:	-1.682000E-01 [C]
T#6 Coef., c_1:	1.007348E+00 [1/C]
Cal Date:	2/23/2005

T#5 Coef., d_0:	1.813000E-02 [C]
T#5 Coef., d_1:	1.000810E+00 [1/C]
Cal Date:	10/7/2004
T#3 Coef., e_0:	6.348000E-02 [C]
T#3 Coef., e_1:	9.978479E-01 [1/C]
Cal Date:	10/7/2004

Water Temperature

Read @ #6 [C]	Read @ #3 [C]	Corrected Ave [C]		Read @ #4 [C]	Air Temperature Read @ #5 [C]	Corrected [C]		Barometric Pressure Read [mm of Hg]	Corrected [Pa]		Rel. Humidity Read [%]	Air Density ref 3 in SP-250 [kg/m³]
20.285	20.287	20.286		20.535	20.529	20.574		746.80	99,565		36.1	1.17732
20.277	20.279	20.278		20.727	20.529	20.670		746.60	99,539		36.0	1.17661
20.443	20.446	20.445		20.844	20.858	20.893		746.50	99,525		37.5	1.17534
20.490	20.492	20.491		21.078	21.084	21.123		746.20	99,485		35.9	1.17407
20.627	20.628	20.628		21.205	21.217	21.253		746.20	99,485		34.1	1.17372

Weight Dry / Weight Full / Weight Drained / Scale Reading / Vessel Volume

Weight Dry Read [kg]	Corrected		Weight Full Read [kg]	Corrected		Weight Drained Read [kg]	Corrected		Scale Reading Read [div]	Corrected		Vessel Volume Contained [m³]	Delivered
168.2385	168.3588		545.6682	545.5915		168.3387	168.4591		2.1	0.000172		0.378609	0.378508
168.2385	168.3588		545.3641	545.6873		168.3387	168.4591		-1.5	-0.000123		0.378598	0.378497
168.2385	168.3588		545.5846	545.9079		168.3411	168.4615		1.1	0.000090		0.378619	0.378516
168.2385	168.3588		545.5055	545.8288		168.3389	168.4593		0.6	0.000049		0.378584	0.378483
168.2385	168.3588		545.8632	546.1866		168.3401	168.4605		5.0	0.000410		0.378593	0.378491

Water Density / Thermal Expansion Corrected Prover Volume / Temperature / Water Density / Kin. Visc.

Ref. Temp. [C]		Water Density ref 4 in SP-250 [kg/m³]	Thermal Expansion Corrected Contained [m³]	Delivered [m³]		Contained [gal]	Delivered [gal]	Contained [in³]	Delivered [in³]		Temperature [C]		Water Density Patterson & Morris [kg/m³]	Kin. Visc. CRC & Kell [m²/s]
15.56		998.146	0.37852342	0.37842287		99.9953	99.9687	23098.92	23092.79		15.560		999.015	1.125E-06
		998.148	0.37851248	0.37841193		99.9924	99.9658	23098.25	23092.12		20.426		998.117	9.964E-07
		998.113	0.37853054	0.37842758		99.9972	99.9700	23099.36	23093.07					
		998.103	0.37849456	0.37839381		99.9877	99.9611	23097.16	23091.01					
		998.074	0.37850144	0.37839946		99.9895	99.9626	23097.58	23091.36					
			0.37851249	0.37841113		99.9924	99.9656	23098.25	23092.07					
				0.37840477			99.9640		23091.68					
			39	38		39	38	39	38					
			0.011	0.011		0.011	0.011	0.011	0.011					

Averages, Thermal Expansion Corrected
Averages, Therm. Exp. & Viscosity Corrected
Repeatability [x 10⁻⁶]
Expanded Uncertainty, [%]

Estimation of Uncertainty k= 2.20 V_{eff} = 11 95% confidence, ref. 8 in SP-250

x_i	Value		$1/V · \partial V/\partial x_i$	Value	$1/V · \partial V/\partial x_i$	dx_i for k=1	$1/V · \partial V/\partial x_i · dx_i$ [x10⁶]
h	--- [m³]		$1/V$	3 [1/m³]		8.2E-06 [m³]	21.6
m_f	545.92 [kg]		$1/(m_f - m_e)$	0.00 [1/kg]		4.2E-03 [kg]	11.0
m_e	168.36 [kg]		$-1/(m_f - m_e)$	0.00 [1/kg]		4.2E-03 [kg]	-11.0
ρ_0	1000 [kg/m³]		$-1/\rho_0$	-1E-03 [m³/kg]		2.0E-03 [kg/m³]	-2.0
β_W	2.0E-04 [1/C]		$-(T_W - T_{ref/w})$	-20.4 [C]		3.3E-07 [1/C]	-6.7
β	4.8E-05 [1/C]		$-(T_W - T_{ref})$	-4.9 [C]		7.0E-07 [1/C]	-3.4
T_W	20.4 [C]		$\beta_W f \beta$	2E-04 [1/C]		2.5E-02 [C]	3.8
ρ_{air}	1.18 [kg/m³]		$1/\rho_0$	1E-03 [m³/kg]		1.0E-03 [kg/m³]	1.0
v							-16.9

100 Gallon

NIST Test Number: 836/123456-05-01
Calibration Date: April 26, 2005 by Sherry Sheckels and John Wright

www.ingramcontent.com/pod-product-compliance
Lightning Source LLC
Chambersburg PA
CBHW081758170526
45167CB00008B/3240